Rumble AND RoaR

SOUND AROUND THE WORLD

SUE FLIESS

Illustrated by **KHOA LE**

Millbrook Press / Minneapolis

Sounds
abound
the world around.

Listen.
Hear.
Cup your ear.

Wake
Stir
Whistle, whir

Hearts
Thump

Beat and pump

Snarl
Growl
Screech and howl

Chirp
Hum
Drip. Drop. Drum.

Babble
Swoosh
Roar and WHOOSH!

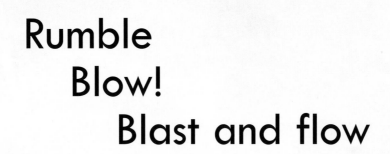

Rumble
Blow!
Blast and flow

Crunch
Snap

Squawk and flap

Crack!
Slide
Mountainside

Trumpet
Pound
Shake the ground

Swat
Swish
    Dive and fish!

Paddle
Slap
Slosh and lap

Tock
Tick
Turn and click.

Whisper
Shush
Slumber, hush.

Purr
Snore
     Sounds . . . no more.

THE SCIENCE OF SOUND

What Is Sound?

The world is full of so many kinds of sounds. Loud, soft, high, low, funny, annoying . . . and more! All sounds start with vibrations (the back-and-forth movements of particles) that travel in the form of waves.

When sound waves reach your ear, they go into your ear canal, hit the eardrum, and cause vibrations. Three tiny bones deep in your ear start moving, and this moves a liquid inside the spiral cavity of the inner ear, the cochlea. The cochlea changes the vibrations into nerve signals that travel to your brain. And your brain makes sense of these signals.

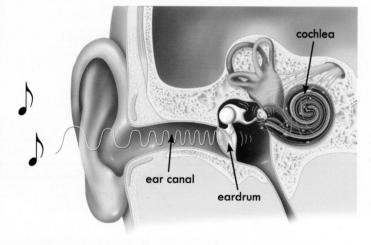

cochlea

ear canal

eardrum

Sound waves move through air, liquid, and solids. On Earth at sea level, sound travels through air at about 761 miles (1,225 km) per hour, which is the same as 1,116 feet (340 m) per second! Amazingly, sound travels more than four times faster in water than it does in air.

FUN FACT
The speed at which sound travels changes depending on the temperature and the type of material the sound is moving through.

Humans can talk thanks to our vocal cords, which are in the larynx, a part of the throat. The vocal cords stretch across the larynx, and they vibrate when air passes through. Hold your fingers gently against your throat and make a sound. Can you feel the vibrations?

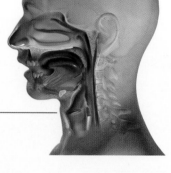

vocal cords open

vocal cords closed

How Does the Heart Beat?

The heart is a special muscle that sends blood all through your body—from the top of your head to the tips of your toes. Think of the heart as a kind of pump. Before each beat, the heart fills with blood. And then it squeezes, or contracts, to push the blood where it needs to go. The right side of the heart receives blood from the body and pumps it to the lungs. The left side of the heart receives blood from the lungs and pumps it to the body. What does blood do? Blood brings oxygen and nutrients to the body parts that need it. It also carries away waste.

Have you ever checked your pulse? Your pulse rate is the number of times your heart beats every minute.

FUN FACT
The heart of a typical seven-year-old who is sitting quietly beats 70 to 110 times per minute!

How Do Animals Use Sound to Communicate?

We might not always understand what animals are saying, but they send and receive many types of messages using sound. They can use sounds for courtship, to find food, to protect their territory, for a display of power, to warn others of a threat, to bond, or to defend themselves from predators.

Many animal sounds are vocalizations—chirps, calls, howls, or even songs. Some click their teeth, drum their beaks, or flap their wings. Many use low-pitched sounds such as rumbles and roars, which tend to travel farther than high-pitched sounds such as squeaks and screams.

There are many interesting examples! Insects, spiders, and fish make sounds by rubbing body parts together. Cockroaches hiss. Bees buzz. Birds croak, grunt, whistle, squawk, and click, and songbirds make musical-sounding notes. Elephants trumpet. Dolphins, porpoises, and whales communicate through whistling, clicking, and forcing air through their blowholes to make chattering or squealing sounds. Ghost crabs beat their claws and drum the sand. Coyotes howl and bark. Hyenas giggle or laugh. Monkeys screech. Donkeys bray. Owls hoot. Tigers roar. Alligators hiss and bellow.

Some animals, such as bats and elephants, communicate with ultrasound or infrasound, which humans cannot hear. Bats use echolocation, also called biological sonar, by producing sound waves into the environment and listening for echoes to return from various objects. This is how they find insect prey. Elephants use low-frequency calls or vibrations to send messages over several miles.

FUN FACT
Crickets chirp faster in warm weather and slower in cool weather. To estimate the temperature in degrees Fahrenheit, count the number of cricket chirps in 15 seconds, and then add 37.

What Makes a Volcano Erupt?

Earth may seem completely solid, but it's not. Under the top layer, the crust, is a layer of liquid rock called magma. A volcano is an opening in Earth's surface, and magma can come out through this opening. Once magma gets to the surface, it's called lava. Lava is incredibly hot when it comes out of the ground. It can reach temperatures of 2,200°F (1,200°C). As it cools, it turns into hard, black rock.

Some volcanoes erupt with a bang, while others erupt quietly. Some eject clouds of hot ash, dust, and lava through an opening called a vent. This ash and dust can cover surfaces in a thick, dark gray powder. They can even be thick enough to block out the sun in the middle of the day. The large rocks hurled from volcanoes are called bombs, and some are as large as a house!

FUN FACT
The world's loudest recorded sound came from the eruption of the Krakatoa volcano in Indonesia in 1883. The roar was heard 3,000 miles (4,828 km) away.

To the Dave Matthews Band, for making the best sounds —S.F.

For D. —K.L.

Millbrook Press™
An imprint of Lerner Publishing Group, Inc.
241 First Avenue North
Minneapolis, MN 55401 USA

For reading levels and more information, look up this title at www.lernerbooks.com.

Additional images on page 30 by BSIP/Universal Images Group/Getty Images (ear); Claus Lunau/Science Source (vocal cords).

Designed by Lindsey Owens.
Main body text set in Tw Cen MT Std.
Typeface provided by Monotype Typography.
The illustrations in this book were created with mixed media and Photoshop.

Library of Congress Cataloging-in-Publication Data

Names: Fliess, Sue, author. | Le, Khoa, 1982– illustrator.
Title: Rumble and roar : sound around the world / Sue Fliess ; illustrated by Khoa Le.
Description: Minneapolis : Millbrook Press, 2022. | Audience: Ages 4–8 | Audience: Grades K–1 | Summary: "The thump of a heartbeat, the chirp of insects, the roar of a waterfall—sound is all around! Rhyming text and atmospheric illustrations present four children encountering all sorts of sounds." —Provided by publisher.
Identifiers: LCCN 2021025103 (print) | LCCN 2021025104 (ebook) | ISBN 9781541598690 | ISBN 9781728445366 (ebook)
Subjects: LCSH: Sound—Juvenile literature.
Classification: LCC QC225.5 .F55 2022 (print) | LCC QC225.5 (ebook) | DDC 534—dc23

LC record available at https://lccn.loc.gov/2021025103
LC ebook record available at https://lccn.loc.gov/2021025104

Manufactured in the United States of America
1-48054-48736-8/6/2021